Rescuing the Disappearing Hedgehog

Toni Bunnell

www.tonibunnell.com

Praise for *The Disappearing Hedgehog* book

'Brilliant. If you are seriously interested in saving these amazing, interesting little guys this book is a must. After rescuing, caring and watching over local hedgehogs since the end of the 2nd World war you would think I had nothing to learn. And so did I; that was until I purchased Toni Bunnell's book 'The Disappearing Hedgehog'. What a delightfully thoroughly enjoyable little book. A really great read to keep handy close by for the advice.' *Gloria Muir, Hogles Wood Hedgehog Home, rehabilitation centre*

'This excellent, clearly written and authoritative book gives precise and detailed information of how to make our gardens more hedgehog friendly, and how to help and look after hedgehogs that are poorly or who have had bad encounters with dogs, foxes, or garden machinery. Very well presented and illustrated. I would thoroughly recommend this book to anyone with an interest in hedgehogs - and buy one for a friend who should be interested but perhaps is not!'
Westfield, hedgehog enthusiast

'I've read my book from cover to cover. Informative and well written.'
Louise Godden, hedgehog enthusiast

'I highly recommend it.'
Geraldine Williams, Veterinary Nurse

'Having 20 years' experience of hedgehog rehabilitation it's good to know the findings and experiences of other rehabilitators. A good read for both the novice and experienced hedgehog carer.'
Leighog

'An excellent book, well presented and factually correct. As an experienced hedgehog carer, I would thoroughly recommend it.'
Susan Haggas, hedgehog rehabilitator

'Excellent little book for anyone with an interest in hedgehogs. Simple and straightforward to read, it contains all the facts you need to know about hedgehogs, plenty of advice and informative illustrations. There are case studies of hedgehogs with a variety of illnesses/injuries from Toni's rescue centre, giving a good overall idea of the problems they can encounter. There is even a test quiz at the end so you can review what you have learned. Handy book to read again and again!'
Amazon customer

'A very straightforward, informative little book. Worth buying to refer to again and again.'
Dawn Higham

The Disappearing Hedgehog book is available from: www.tonibunnell.com

It includes: ecology and behaviour, what to do when you find a hedgehog, exceptional success stories, tales from York Hedgehog Rescue Centre, helping hedgehogs in your garden.

Also by Toni Bunnell

Music Makes a Difference[1]

The Room Between the Floorboards[2]

Samuel and the Stolen Words[2]

A Door in Time[3]

The Fidgit[2]

A Life Well Lived[2]

The Disappearing Hedgehog[1]

The Nameless Children[2]

The Dead Space[1]

The Dark Mirror[1]

Trapped[1]

***

Available as paperback only[1]

Paperback and ebook[2]

Ebook[3]

www.tonibunnell.com

This book is dedicated to Lorraine Jackson, friend and fellow hedgehog carer, who has been running Hull Hedgehog Rescue for the past twenty years.

Contents

Introduction	1
Reasons for decline	3
Incidence of disease and injury	6
Basic equipment	10
Assessment, treatment and outcome	13
Ailments: diagnosis and treatment	37
Breeding season	43
Early and late litters	45
Mums with babies	50
Caring for hoglets	59
Sight impaired hedgehogs	67
Why do ticks target hedgehogs?	73
Hibernation	77
Weight/size relationship: A new index	80
The borderline hedgehog	83
Post-release monitoring	86
Means of identification	87
Publications	90
The Author	94

First published in 2016 by Toni Bunnell, York, UK

www.tonibunnell.com

ISBN: 978-1-78280-890-9

All rights reserved. No part of this publication may be reproduced, stored in or introduced into a retrieval system, or transmitted in any form, by any means (electronic, mechanical, photocopying, recording or otherwise) without prior written permission from the copyright holder Toni Bunnell.

A CIP record for this book is available from the British Library.

Printed and bound in Great Britain by
Short Run Press Ltd Exeter

This book is sold subject to the condition that it shall not, by way of trade or otherwise, be lent, resold, hired out, or otherwise circulated without the author's prior consent in any form of binding or cover other than that in which it is published and without a similar condition, including this condition, being imposed on the subsequent purchaser.

Cover / text design and photographs by Toni Bunnell

www.tonibunnell.com

Introduction

Since I wrote The Disappearing Hedgehog book, in 2014, the European hedgehog has shown no signs of recovering from its steady downward spiral in numbers in the UK.

The past fifty years has seen a 97% reduction in numbers, from around 32 million to less than one million. It is not just the fall in population density that is causing concern, but the rapidity at which this has occurred. The current decline points to the imminent extinction of the species in the UK.

Whether the hedgehog will vanish completely from our shores, or whether populations will continue to exist in fragmented areas, as has happened with the bobcat in the US, it is impossible to know. It is safe to say, however, that faced with the rate of decline, the future for hedgehogs is not looking good.

The reasons for the rapid decline in hedgehog numbers are examined at the beginning of the book. Suffice it to say that humans play a not insignificant part in the downfall of this iconic animal. This book also explores some of the ways in which we, as individuals or in groups, can help to redress the balance.

As a wildlife biologist, with an interest in mammal conservation, I first became alerted to the problems being experienced by hedgehogs in the UK, in 1990. It was then that I set up York Hedgehog Rescue, with the aim of helping to rehabilitate injured and sick hedgehogs and release them back to the wild. It has been an uphill struggle even though my success rate is high.

Hundreds of hedgehogs have arrived for treatment at my rescue centre over the past 28 years. Many of them are juveniles who arrive in the autumn, cold and hungry. I weigh and assess them, then make them cosy in a warm unit. Most of them need medication for their various ailments. Those that are injured go straight to the vet. Hedgehogs arrive in all shapes and sizes and I endeavour to make their lives, while they are with me, as comfortable as possible.

In this book I share the findings of my research, resulting from analysis of data I have collected. I also look at ways in which all of us, whether members of the public, or experienced carers, can help hedgehogs in need. Hopefully, this will help to reverse the trend and hedgehog numbers will begin to increase.

Reasons for decline

In 2011 I was approached by a major television company and asked to produce a report on endangered species in the UK, listing the ten that were most endangered. Based on a review of the literature I placed the hedgehog at number ten, due to their numbers having fallen from over 32 million to less than one million in the space of only fifty years. These figures are supported by every study that has attempted to determine population levels of hedgehogs throughout the UK.

Of particular concern is not just the fall in numbers, but the rate of decline. At 97% this signifies critical endangerment leading to extinction. The reasons for this are well documented and include habitat loss, and the hazards presented by humans in suburbia. Other possible culprits include pesticides and infections transmitted by as yet unknown pathogens.

The reasons why hedgehogs are being presented at wildlife rescue centres, in need of care and treatment, are many. Indeed, the whole issue facing the hedgehog is multi-faceted. High on the list is habitat loss, fragmentation of habitat, and the reduction of green corridors that allow small mammals, such as the hedgehog, to move safely from one area to another, as well as providing nesting sites. The countryside, previously the habitat of choice for the hedgehog, has now been replaced by suburban gardens. These, however, bring their own problems.

In recent times the decimation of gardens which occurs during their redesign, termed 'garden makeovers', has seen the removal of many ideal sites where hedgehogs could forage, nest and rear their young. This is not helped by an increased tendency to erect garden boundaries that prevent movement of hedgehogs from one garden to another. This is easily remedied by making a gap at several places round the perimeter of the garden. Hedgehogs thrive in gardens where open access is available, a gap only 10cm high and wide being adequate for a hedgehog to enter and leave a garden, preferably at three or more different sites.

Other problems that hedgehogs have to contend with include dog attacks, strimmer injuries, drowning in garden ponds, becoming entangled in netting, ingesting rat poison, eating slugs containing slug pellets and becoming trapped in holes in the ground. All these hazards are avoidable, with a little care and consideration, and are discussed in The Disappearing Hedgehog book.

Another issue that is now receiving attention is the massive fall in invertebrate numbers, 45% over the past thirty five years. This could be a contributory factor where the rapid decline in hedgehog populations is concerned. As the diet of hedgehogs is comprised of 30% beetles, 25% caterpillars, 5% millipedes and 5% slugs and snails, the reduction in invertebrates, other than slugs and snails, will have serious consequences for hedgehogs.

Anecdotally, the spring of 2015 saw many rehabilitated hedgehogs, that had been released fit and well from April onwards, being returned to rescue centres for treatment, having been found out in the day. On assessment these animals were found to be malnourished and carrying a higher than acceptable burden of the internal parasite, the lungworm.

Hedgehog rehabilitators throughout the UK, myself included, suggested the possibility that the lack of beetles etc had resulted in hedgehogs eating a higher amount of slugs and snails than they would normally do, these being the intermediate host for the lungworm. This would increase the chances of hedgehogs becoming infested by lungworm and decrease their chances of survival.

It is only by looking at the many confounding factors affecting hedgehogs, that we will have any chance of reversing the current trend that suggests a downward spiral to extinction. It is possible that isolated, viable, populations will continue to exist in various parts of the UK. However, these will likely not be known to us and, to all intents and purposes, the hedgehog will have disappeared from our lives.

Incidence of disease and injury

In order to shed further light on the reasons for population decline in the British hedgehog, and in the hope that any findings might be used to help reverse this trend, I conducted a study (Bunnell, 2001).

Data were collected to determine the number of hedgehogs that were heavily burdened with parasites, suffered myiasis (flystrike), leg breaks, eye infections, hypothermia and other ailments, and were hit by cars. The data were then used to determine the impact of these ailments and injuries on the survival rates of hedgehogs presented at York Hedgehog Rescue over a three year period.

The main reasons why hedgehogs were brought to my rescue centre, in York, between 1998 and 2000 were assessed. Of the 168 arrivals, 61% were nestlings, 25% juveniles and 14% adults. Significantly more male nestlings arrived than females, with a male-biased sex ratio (male:female = 3:2). There were also more males than females across all age categories combined.

The vast majority (94%) of the animals were still alive 48 hours after arrival. All animals were released to the wild as soon as they were restored to full health. During the 20 days following their admission, of the 168 arrivals 34 died, giving a survival rate of 84%.

Combining the data across the age categories, the most common ailments were malnutrition, which affected 27% of animals, and dehydration (15%).

Nestlings were particularly affected by malnutrition, accounting for 55% of those affected, and dehydration (52%).

Puncture wounds were evident in 5% of animals, mainly being sustained from dog bites.

Myiasis (flystrike) was seen in 2% of animals, all nestlings. In each instance the animal was also suffering from hypothermia.

The hedgehog tick (*Ixodes hexagonus*) was present in 14% of hedgehogs (63% of these being juveniles). As the infested animals were all casualties requiring treatment, I considered that it might be expected that healthy wild hedgehogs would be less likely to carry ticks.

This suggestion was supported by data collected over ten years, which indicated that the number of ticks infesting a casualty seemed to be directly related to its state of health and prognosis for a return to full health.

The theory was tested and proven during a later study that I carried out (Bunnell, 2011).

Those animals possessing more than 25 ticks mostly failed to survive. The greatest number of ticks recorded on an individual animal, between 1998 and

2000, was 72. However, in 1997, I treated a hedgehog bearing 199 ticks, an excessive number for such a small mammal.

Ringworm, a fungal skin condition, was found in 4% of all animals, though mainly in nestlings.

Sarcoptic mange was observed in 6% of animals and was sometimes accompanied by ringworm. This is not surprising as the mites responsible for the mange could also help to spread the ringworm, as the fungus has been isolated from their droppings.

Demodectic mange was observed in only 2% of animals.

The nematode *Crenosoma striatum* (lungworm) was the parasite most commonly noted to be present in excessive numbers, with an accompanying change in the gross morphology and texture of the faeces (11% of total) (Bunnell 2001b).

The tapeworm, *Hymenolepis erinacei*, was present in excessive numbers to the point of producing changes in the faeces, coughing and weight loss, in 5% of all animals.

Broken limbs, thought to have been caused by dogs, were seen in two animals only (1%), while miscellaneous injuries caused by human-related activity were observed in 4% of all animals.

Nestlings displayed the highest survival rate at 74%, with juveniles and adults surviving at the almost equal rates of 55% and 58% respectively.

In summary: The most common ailments were malnutrition (27%) and dehydration (15%), while ticks (14%), lungworm (11%), ringworm (4%) and sarcoptic mange (6%) were also found. Road accidents and other injuries each affected 4% of the arrivals.

Shallow, heavy dishes for food and water

Two types of weighing scales

Covered hot water bottle for emergencies or when travelling

Basic equipment

- Hutch: preferably minimum 4 ft x 2 ft
- Vet recovery heat pad
- Shallow, heavy dishes for food and water
- Newspapers for covering bottom of cage
- Small pet carrier for transportation
- 1 ml plastic syringes for administering medicine or syringe feeding
- Towels without torn edges (strangulation risk)
- Weighing scales
- Record keeping system
- Gloves for handling
- Tweezers for removing ticks
- Flea powder free from Pyrethrum
- Malaseb shampoo for bacterial skin infections

Medicines

- Baytril (oral antibiotic) or similar
- Telmin (lungworm)
- Panacur (tapeworm)
- Septrin (acts against the protozoan responsible for coccidiosis and also acts as an antibiotic)
- Tea Tree Antiseptic Cream (ringworm and sarcoptic mange)
- Neem Oil (ringworm and sarcoptic mange)

Gloves for handling

Syringe for feeding and administering medicine

Tea light useful to show size in a photo

Assessment, treatment and outcome

To give an indication of the procedure I follow, at York Hedgehog Rescue, I am going to outline the progress of ten hedgehogs, from the moment they arrived with me.

Bilbo

On January 8, 2016, a small male juvenile was found during the day in a back garden of a cul-de-sac in York. I collected him, named him Bilbo, and charted his progress. At 275g he must have been born to a late litter towards the end of 2015. He was underweight and was placed on a heat pad. The vet diagnosed pneumonia and prescribed a course of the oral antibiotic, Baytril, as well as Telmin for lungworm and Septrin for coccidiosis.

Over consecutive days his weight (g) was as follows: 274, 282, 304, 318, 351, 350, 340, 341, 350, 368, 381, 402, 406, 396, 414, 431, 457, 462, 485, 483, 486, 520, 533, 546, 542, 583, 593, 618, 574, 631, 648, 678, 740, 785, 843, 834. On February 26 he weighed 883g and two days later 895g. On March 5 his weight had fallen to 853g and another course of Telmin was prescribed to treat lungworm. Over the next three days he gained 33g (886g) and by March 11 had reached 957g.

By March 25 he was 1043g and was released back to the garden where he had been found 76 days earlier.

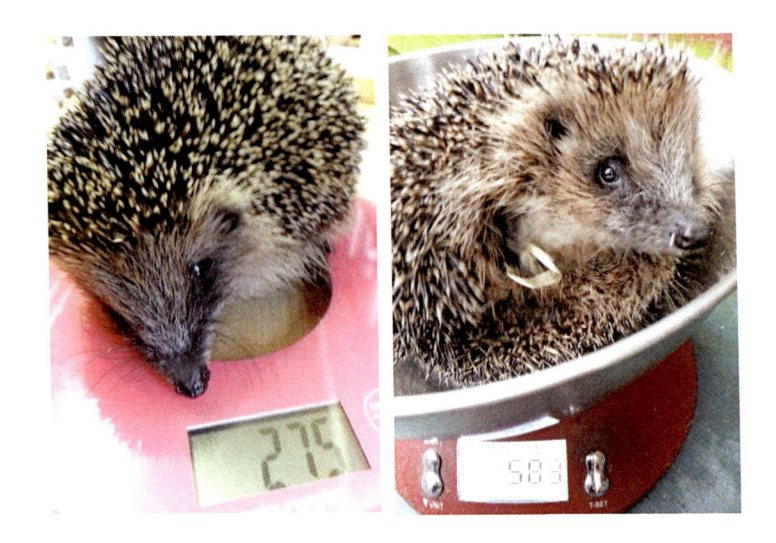

Bilbo on arrival   26 days after arrival

Bilbo free to roam my garden while gaining weight

Squidge

Squidge arrived on July 6, 2015, after being found out in the day in a green area in the middle of York. He was placed on a heat pad in the kitchen as he weighed only 99g. To begin with he progressed nicely, his weight (g) being recorded as 100, 107, 111, 112, 117, 115, 126, 127.

On July 16 two ticks were found on him and removed. Also, his weight had fallen by 3g, which was 2.3% of his total body weight, and so needed to be acted upon. A second course of Baytril was started. He continued to gain weight (g): 138, 133, 144, 156, 163, 177, 171, 186, 196, 200, 207, 238, 240, 232, 248, 261, 293, 307, 331, 329, 350, 375, 390, 431, 442, 530, 544.

By now he had been moved to the large (8ft x 3ft) two-tiered outside enclosure, where he had more space to wander but could still be found easily and weighed each morning. On August 28 he weighed 565g and was nearing the time when he could be released. Three days later, without warning, I found him dead in his nest. There had been no sign that he was poorly and the discovery was quite a shock.

In the 26 years I have been caring for hedgehogs I have only once come across a similar incident. A hedgehog that had reached 350g, and was roaming freely in my garden while he gained weight, was found dead one morning in the middle of the lawn. The post mortem revealed a large thoracic tumour that had suddenly haemorrhaged. I have included this story to show that sometimes, despite out best efforts, not all hedgehogs in our care can be saved and safely returned to the wild. We can only ever do our best.

Left: Squidge on arrival
Above: 126g, tick over left eye

Below: Squidge at 390g. Tea light for size comparison.

Left: Squidge at 530g having gained 88g in nine days

Below: Buster on his release day showing his unique emulsion mark and blue tag

Buster

Buster was found on April 7, 2015, in the green area that I am monitoring in the middle of York, within the city walls. I collected him and assessed and weighed him. At 462g he appeared 'light for size'. He was not injured and the poo was normal so no treatment was administered at that point.

His weight (g) was recorded daily to begin with: 462, 489, 499, 509, 515. On April 14 he weighed 529g, followed by 540g the next day, and 545g two days later. After this his weight was rather up and down: 548, 524, 572, 557, 568, 513, 505, 523, 535, 542, 560, 549, 559, 573, 610, 594, 623, 616, 606.

On June 3 Buster weighed 644g, a gain of 38g from two days previously. However, he had a rattly chest which is symptomatic of lungworm. According to my vet, a rattly chest can signify old lung damage and not warrant treatment. This proved to be the case and Buster continued to gain weight and improve. His weight record (g) for the following days: 636, 628, 664, 628, 639, 636, 679, 690, 704, 720, 777.

On July 7, 91 days after he had been found out in the day, Buster was returned to the same spot where he had been found. He continued to be picked up by the Infra-Red Reconyx cameras, his unique emulsion mark being easily detected. He was also marked with a tag to aid identification, just in case he got into difficulties once more and the emulsion had worn off.

Hoggle

On April 16, 2016, Hoggle was found lying by the roadside, not moving. He was taken to the Minster Veterinary Practice, York, where they removed all the fly eggs and treated him with Xenospot. Two days later I collected him from the vet and his progress was as follows:

Weighed in at 707g. His weight fell to 701g the following day and he was prescribed a course of Telmin (lungworm) and Septrin (coccidiosis). This treatment regime took effect immediately and Hoggle's weight rocketed to 774g the next day. Subsequent daily weights (g): 748, 787, 812, 856, 879, 887, 888, 889, 903, 955, 969, 990, 1002, 1072, 1084, 1090, 1095, 1095, 1061, 1032, 1009, 1053, 1029, 1055, 1066, 997, 977, 1014, 1042, 1057, 1030, 1018, 1044, 1047, 1027, 1023, 1090, 1057, 1050, 1007, 1018, 1023.

Having done so well for two months Hoggle's weight suddenly crashed from 1023g to 952g, a loss of 71g. He also had a sore area above one ear. I took him straight to the vet where they examined him under light sedation, in order to check him over. He was found to have an extensive patch of moist dermatitis underneath. The outcome was not good. As the area affected by moist dermatitis was badly infected they felt that the only course of action was to put him to sleep. This was a very sad end to what had been a great success story for an old hedgehog approaching the end of his life, and who was going to live out the rest of his days in safety at my rescue centre.

Hoggle, two months after arrival and doing well

Zola 3 days after arrival    Zola 11 days after arrival

Zola

On November 4, 2015, a little female hoglet was taken to York RSPCA having been found out in the day. I collected her, placed her in the 'intensive care' area in my kitchen, with an optional heat pad that she could move on or off as she wished. She weighed 274g, had most likely become separated from her mother and siblings during a night out foraging and had since become poorly, hence her appearance in the day.

I gave her 15ml subcutaneous fluids and she was prescribed a course of oral Baytril (antibiotic) and Telmin (lungworm), in addition to Septrin, due to the presence of coccidiosis, as indicated by the poo. She responded well to treatment, her weights (g) being: 274, 270, 263, 298, 313, 336, 361, 374, 391, 393, 392, 410, 409, 420, 414, 451, 448, 469, 451, 482, 506.

When she reached 451g she was moved to my heated unit with a thermostatically controlled radiator, and an optional heat pad in her house. On November 27, at 506g, she was transferred to my insulated shed. Her record continues: 502, 524, 544, 558, 566, 570, 589, 600, 615, 631, 651, 644.

On December 16, when she had reached 644g, I moved her from her 4ft hutch to my enclosed garden where she could continue to gain weight while being monitored. Her subsequent weights (g): 617, 642, 634, 640, 636, 623 (two ticks removed), 635.

Above: Zola, 615g, 12/12/15

Right: Zola just out of hibernation, 473g, 27/03/16

Zola, before hibernating. December 27, 623g

Zola's weight then fell from 635g to 608g and she was put back in the insulated shed where it fell further to 576g. The sudden development of a rattly chest indicated lungworm and a course of Telmin was prescribed. She was moved back to the heated unit and began to hibernate on January 16, at 563g.

On February 2, sixteen days later, she was still hibernating in the heated unit with unlimited food and water available. She weighed 519g which indicated the standard rate for weight loss during hibernation: 44g in 16 days = 2.75 g/day. She continued to hibernate and was weighed intermittently, being 495g on February 16. She finally woke from hibernation on March 27, weighing 463g.

After breaking hibernation Zola gained weight rapidly, reaching 705g on May 9 when she was released to a suitable site where she would be monitored by my cameras. Release back to where she was found was not possible as the finder had failed to leave their contact details with the RSPCA.

Samuel on arrival with ringworm on his nose

Samuel responding well to Tea Tree Antiseptic Cream

Samuel shortly before release with nose fully healed

Samuel

On August 26, 2014, Samuel was found out in the day in a green area in York by one of the gardeners, who notified me. I collected him immediately. He was then weighed and assessed. At 486g he was 'light for size' but other than that and some ringworm on his nose, the usual starting point for ringworm, he appeared to be fine.

The ringworm was treated using Tea Tree Antiseptic Cream, which also works well for sarcoptic mange.

Samuel had a healthy appetite and gained weight quickly for the first three days: 490, 499, 508g. His weight then fell to 504g and he was prescribed a course of Telmin as he was found to have lungworm, common in juveniles at this time of year. This did the trick and he continued to gain weight (g): 525, 552, 567, 606.

By September 3, only eight days after being admitted, he had gained 120g, 39g overnight on the final day. He now possessed the correct weight/size relationship and was released back where he was found, complete with markings and tag.

Polo

On July 6, 2015, Polo was found in the green area that I am monitoring in York. She was found by Martin Ball, the head gardener, who is very aware that hedgehogs should not be out in the day and always takes the appropriate action. He also goes to great lengths to ensure that the hedgehogs in the area are not exposed to any hazards such as strimming or pesticides. As a result the area is well populated by hedgehogs that breed and survive from one year to the next.

Polo was a typical underweight, 'light for size' juvenile who weighed 435g on arrival. I housed her in a large hutch in my insulated shed and observed her. Although she gained 4g on the first night, she was found to have lungworm and coccidiosis and was treated with Telmin and Septrin respectively. The response was immediate as demonstrated by an increased appetite and weight (g) gain: 478, 563, 556, 673, 703.

On July 20, fourteen days after being found, Polo was released back to the same area, having gained 268g since her arrival. She was marked with a unique emulsion mark and two brown tags.

Pumpkin

Pumpkin came to me from York RSPCA on October 26, 2014, weighing 285g. He had a head injury and his right eye lid was very swollen. He was also circling rapidly and was disorientated.

York RSPCA staff took him to Minster Veterinary Practice who treated him and gave him subcutaneous fluids. He was also given Metacam as I thought that he might have been pecked by a bird and was most likely in pain. I collected him on October 30.

When he came to me I removed five ticks and placed him in my heated unit. Initially he was very hyperactive. He was prescibed Telmin for lungworm as he had a rattly chest, and a course of Septrin for coccidiosis. This was very effective and Pumpkin gained weight rapidly. He was then moved to my insulated shed.

On November 25, he began to hibernate at 524g. Six days later he had lost 20g which at, 3.2g/day, is an acceptable rate. He emerged from hibernation on December 14 at 468g, then went back to sleep until December 23 when he finally came out of hibernation weighing 448g. I moved him to the heated unit and he was prescribed another course of Telmin as his chest was rattling. He continued to do well and his weight increased until it reached 500g.

Pumpkin was allowed to gain weight in the outside enclosure, to give him room to roam. On February 16, 2015, at 838g, he was released to a safe area which I am monitoring.

He was marked using emulsion paint and a plastic tag. An identical emulsion mark was placed on each side so that he could be identified by the camera images, regardless of which side was presented to the camera as he walked past.

Eight days later he was found out in the day not far from where he had been released, weighing 623g, 215g less than his release weight. I collected him and monitored him. He did not require medication, only food.

Pumpkin rapidly gained weight and on March 25 he was released once more to the same area, weighing 857g. He was not seen out in the day again which was taken as a good sign.

Pumpkin reaches 500g after emerging from hibernation

Pumpkin 503g
York Hedgehog Rescue © Toni Bunnell 2015

Unique emulsion mark on left and right sides plus yellow tag
Below: Pumpkin on final release day

Lucky

On July 20, 2014, I received an email from the security officer of one of the York Colleges to say that a hedgehog had been found in a box in student accommodation. The student had left the country without informing anyone of the existence of the hedgehog. I named him Lucky as it was only due to the astuteness of the security officer, who heard a scrabbling sound in a cardboard box in the room where he was discovered, that he didn't starve to death.

I collected the little hedgehog, who weighed only 233g and had an injured hind leg which he was dragging slightly. He was also very thin. I took him to Minster Veterinary Practice where they X-rayed him and discovered that he had an old fracture to the leg that appeared to have healed.

Eighteen days after arrival Lucky started to lose weight. He had done well to start with, reaching 437g, and was being monitored in my garden prior to release. However a slight chest rattle, accompanied by the weight loss, indicated lungworm and a course of Telmin was prescribed.

He rallied and was moved back to my enclosed garden on September 20, weighing 554g. His weight then began to fall again and he was prescribed Telmin, Septrin and Baytril, following a visit to the vet, in addition to Profender Spot-on for fluke.

He then became progressively more poorly and was diagnosed with an infection, the source of which could not be identified but might have been contracted from birds that frequent the garden where he was roaming free.

As he had no appetite I had to syringe feed him Hill's Prescription Diet a/d for two weeks until he began to eat on his own. On October 21, weighing only 396g, Lucky ate a few mealworms, an appetising treat that should only be fed in moderation to hedgehogs. This is due to the lack of minerals and vitamins and the tendency for them to be addictive to the point that hedgehogs refuse to eat anything else but mealworms!

Having begun to rally, he once again started to lose weight and was experiencing serious difficulties walking, particularly with moving his back legs.

I started to syringe feed him Hill's a/d once more and continued with this for five weeks. He also received Ivermec injections from the vet during this time, in the hope that this would help to rectify the neurological condition that he had acquired.

From November 19 to December 25 I syringe fed him several times a day and it paid off! Eventually, he began to walk a few steps at a time, to eat on his own and to gain weight. The lengthy treatment and care he received was all worth it.

On February 8, 2015, Lucky had reached 880g and was moved to the outside enclosure. At the beginning of March he began to hibernate intermittently, finally emerging from hibernation on April 10, 2015, weighing 634g.

On October 6, 2015 he started to hibernate at 786g. On December 14 he broke hibernation at 625g.

He has continued to live in my enclosed garden as he has not proved to be a good candidate to be released back to the wild.

Lucky on his finding day. Malnourished and weighing only 233g he was taken to the vet who diagnosed an old fracture to one of his back legs.

Lucky, having fully recovered from all his ailments, prepares to walk up the ramp to the upper nest box in the outdoor enclosure. Here he weighs 880g.

Left: Dottie 3 weeks after arrival at 276g
Below: Blue and brown tags

Dottie shortly before being released to the wild

Dottie returning for care at 419g and infested with ticks

Dottie

Dottie came to me as a small hoglet, weighing 265g, from York RSPCA on October 3, 2015. She was prescribed Septrin by the vet for coccidiosis and responded well to treatment. She was moved from my kitchen to the heated unit, then the insulated shed, followed by the outside enclosure on November 23.

On December 4 she was released to my enclosed garden to spend the winter. Even though, at 677g, she had reached my recommended minimum pre-hibernation weight of 650g, I did not plan to release her until the spring as the weather was set to change for the worse and it wouldn't have taken much for her to slip below 650g.

On May 18, 2016, Dottie was released to a safe area in York that I am monitoring. She weighed 812g. She was marked and tagged with a blue and brown tag.

On July 11, 54 days later, she was found lying in the open by Martin Ball, the head gardener of the estate. I collected her immediately. She was found to have six ticks and was very thin, weighing a mere 419g. The decrease of 393g amounted to a 48% weight loss since her release.

I removed the ticks and she was prescribed a course of Baytril, an antibiotic, to address a bacterial infection as well as Telmin to treat lungworm. Her appetite was poor so I began to syringe feed her and administered subcutaneous fluids that I have been trained to do by a vet at Minster Veterinary Practice.

When syringe feeding I held her upright so that no liquid flowed back into her trachea which could cause inhalation pneumonia. She was still wobbly the following morning so I gave her some more subcutaneous fluids and syringe fed her again. She had gained 12g overnight and was eating a little on her own but I was not prepared to take any chances.

Very sadly, despite my best efforts, Dottie died in the evening on July 18, only seven days after being readmitted. It was a very sobering reminder that, however hard we try to save the hedgehogs in our care, at the end of the day they have to fend for themselves in the wild.

In this instance Dottie would appear to have succumbed to the effects of lungworm. Lungworm is a major problem for wild hedgehogs and appears to be an increasing hazard as the amount of invertebrate food, other than slugs and snails, becomes seriously diminished in their habitat.

Ailments: diagnosis and treatment

Moist dermatitis

Several years ago, a hedgehog that I was caring for developed red, sore areas on the top of both front feet. She had been free to roam in my garden and the vet suggested that she had trodden in something caustic. This was not possible as the garden is, naturally, wholly hedgehog friendly. The nature of the skin condition was never identified.

In 2016, a very old hedgehog, Hoggle, had been nurtured back to health and was being monitored in my garden. She suddenly developed red areas on top of her front feet and was rushed to the vet. In this instance the condition was correctly diagnosed as moist dermatitis. It had spread to the entire underside of the hedgehog, this only becoming apparent following light sedation by the vet.

Moist dermatitis is a superficial skin infection that occurs when normal skin bacteria overrun the skin's natural defences. It can occur when the skin's surface is damaged, and is exacerbated by damp bedding. Another hedgehog, Beacon, also developed moist dermatitis during a hot, humid period in 2016. He was treated by bathing the affected area daily using Malaseb shampoo. This has the effect of killing the bacteria while allowing the skin to heal itself. It is a very effective treatment and Beacon fully recovered.

Beacon showing skin on right shoulder badly affected

Beacon being bathed with Malaseb shampoo

Skin completely healed following treatment

Dragging leg / Amputees

A reluctance to place a foot on the ground, to bear weight, suggests an injury of some sort. Fractured legs can sometimes heal on their own, providing there is no infection in the bone. A vet will confirm whether a break has healed and whether the leg is capable of being saved.

If amputation is necessary, the general consensus is that a hedgehog without a front leg is not going to have a good quality of life and that it is kinder to euthenase. This is because the remaining stump will drag on the ground and become sore and infected.

When a hind leg has to be amputated the hedgehog, if otherwise in good health, will be able to manage extremely well. Return to the wild is not deemed to be acceptable, but being able to roam freely in a supported, enclosed garden, is a very viable alternative. Hedgehogs minus a hind leg are able to move at speed, mate, and the females have babies.

Sometimes, a leg can be dragging but not broken. In the case of Frankie an X-ray showed no break, but rather an injury to the shoulder. This rendered him unable to turn his foot out in order to place it flat on the ground. Consequently, a sore developed as he was forced to walk on his knuckle (see photo). Frankie was otherwise healthy but was confined to limit further damage to his foot and allow it to heal.

In the space of two weeks the injury to his knuckle had healed and he was placing his foot flat on the ground. Observation of him roaming freely in my garden confirmed that his leg was functioning properly. He was released to the garden where he had been found and where he is a regular visitor to the feeding station.

Frankie walking on his knuckle

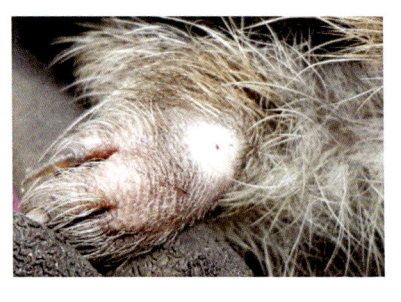

Very sore foot   Foot completely healed

Flystrike

Flystrike, otherwise known as myiasis, is often difficult to diagnose when observed for the first time. Sometimes described as tiny grains of rice, it can appear as a solid mass. It is crucial to remove every single egg as they quickly hatch out into maggots which will then begin to devour the flesh of the hedgehog and enter orifices such as the eyes, anus and mouth. The outcome for the hedgehog is then bleak. The eggs can be removed using tweezers or a tooth brush. The important thing is to remove them all.

Ticks

Sick hedgehogs are often infested with ectoparasites such as ticks. Tweezers are useful for removing ticks as well as fly eggs and fleas. It is important to grasp the tick as close as possible to the point where it is embedded in the hedgehog. A sharp tug towards you, or an anti-clockwise twisting motion, will then serve to remove the tick without leaving the mouthparts in the hedgehog.

Ringworm

This skin condition is caused by a fungus. It can be treated by applying Tea Tree Antiseptic Cream, or Neem Oil, sparingly to the affected area, daily for a week.

Myiasis (flystrike)

Tick removal

Ringworm often begins on the nose

Tea Tree Antiseptic cream smeared on nose

Ringworm healed following treatment

Breeding season

Hedgehogs are purported to be solitary animals that rarely interact, the exception being during courtship and mating. I have not found this to be the case. While checking the area I am monitoring with cameras, I have come across hedgehogs in close proximity to each other. In one instance four hedgehogs were foraging together at dusk underneath some trees. They were all adults and not a family group.

Sharing a nest is also a common occurrence, however, defending a food source against all comers is often witnessed.

With so many hoglets arriving at my rescue centre over the years, I decided to analyse data I had collected. I was curious to know how long the breeding season lasted, and whether there was more than one period when breeding took place.

With the exception of a solitary animal in December and January, young hedgehogs (up to 250g) arrived at my rescue centre between June and November.

My data confirmed that European hedgehogs start to produce litters from June onwards, over a five month period, with the greater majority of young being born in June. Two distinct peaks, the first in June and the second in October, suggest two breeding periods (Bunnell 2001) or, at the very least, an extended breeding season.

It is not known whether young born during the second breeding peak result from the first litter born that year to a particular female, or from a second litter from that animal.

However, other researchers have confirmed that when conditions are favourable hedgehogs can raise two litters in a season. Indeed, on the North Island of New Zealand, where the climate is mild, the introduced European hedgehog is thought to commonly have three litters a year.

Jackson (2006) reported that, in Scotland, at least 96% of adult females attempted to breed early in the season, resulting in litters born in June, while the greater majority (81%) bred again later in the season, resulting in litters born after mid-July.

Early and late litters

Having established that there were two distinct breeding periods in the UK I wanted to shed more light on the survival chances of hedgehogs born to late litters. Assumptions had previously been made that second, or late, litters would be very unlikely to survive the winter.

For my study early litters were defined as all young hedgehogs, weighing no more than 255g, that arrived at my rescue centre between June and the end of August in the same year. Late litters were defined as all young hedgehogs, weighing no more than 255g, that arrived between the beginning of September and the end of January the following year.

Figure 1 Young born to early and late litters (1998 – 2006)

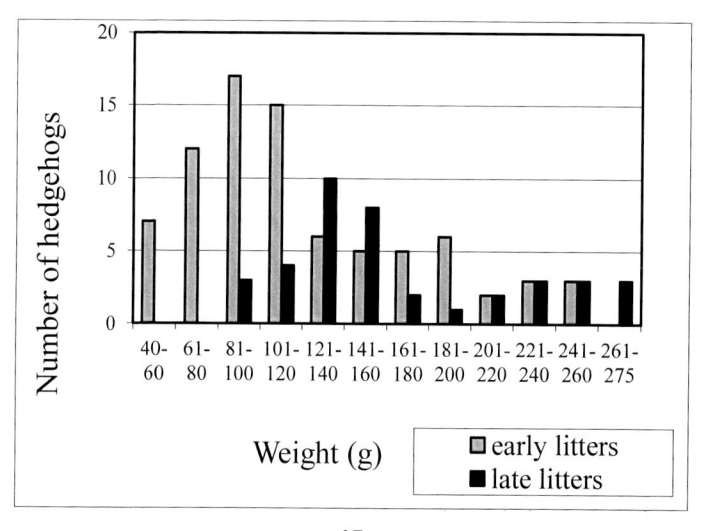

Figure 1 illustrates the number of hedgehogs born to early and late litters and their weights.

I put forward the theory that young European hedgehogs born in the autumn gain weight more rapidly than their counterparts born early in the season. This, in turn, might serve to increase the survival probability of late litters, if survival probability is improved by a relatively high growth rate. It has been suggested by other researchers that growth rate is adaptively flexible and affects optimal size and development times in a seasonal environment, the most likely effect being that growth rate becomes faster with less time available.

I analysed data collected over nine years (1998 – 2006) (Bunnell, 2009).

The daily growth rate (g/day) was determined for all healthy young hedgehogs (n=87) born to early and late litters between 1998 and 2006. This enabled the overall rate of growth from time of arrival of each hedgehog to maturity or release status to be determined.

The overall growth rate of young born to late litters was found to be significantly higher than those born to early litters (Table 1). There was no significant difference in growth rate between the sexes.

Table 1 Mean values for overall growth rates in early and late litters

	Early litters	Late litters
Mean value (g/day)	10.87	12.55
No. of animals	61	26

Table 2 Mean number of days taken to reach 650g

Early litters	Late litters
59.8	51.8

Table 2 shows the mean number of days taken for hedgehogs from late litters to reach 650g, the minimum pre-hibernation weight that I recommend (Bunnell, 2002). On average, it takes eight days less for young from a late litter to reach 650g, compared with young from an early litter.

Regardless of whether a late litter is the first or second born to any particular female, it could be speculated that young hedgehogs born to late litters have a reduced chance of surviving the winter. However, if all the young were to succumb during the first cold spell of winter this would mean a great waste of resources on the part of the female hedgehog giving birth to late litters.

In terms of conserving resources, it would seem more beneficial to an adult female hedgehog not to produce a late litter. She would be less likely to survive hibernation if she has had to sacrifice some of her body reserves to feed her young. If the young then fail to survive it will all have been to no avail.

This study has shown that, in northern England, European hedgehogs born to late litters gain weight significantly faster than those born to early litters. Therefore, it is not the case that late litters automatically have a poor chance of survival.

The reasons for the significantly faster growth rate seen in juveniles from late litters can be speculated upon. It is possible that decreased day length, accompanied by falling ambient temperatures, serves to alter the physiology of hedgehogs born later in the year. An increase in appetite, supposing sufficient food supplies to be available, would result in a more rapid weight gain in late juveniles compared with their counterparts born to early litters.

As the hedgehog in Britain is able to forage and obtain its own food when it has a body weight of 200g, the ability to gain weight depends on the efforts of individual animals and the availability of food, in terms of quantity and quality, and not solely on parental care. In New Zealand, on the North Island, European hedgehogs have been observed foraging at a weight of 100g (Bunnell, personal observation, 2003).

The tendency of young from late litters to gain weight at a faster rate than young from early litters increases the chance of late juveniles reaching an acceptable weight prior to hibernation. I established the minimum pre-hibernation weight (coupled with a rounded end) to be 650g (Bunnell, 2002).

These findings dispel previous suggestions that all young hedgehogs born late in the year are automatically doomed to die due to a failure to achieve a satisfactory weight which would allow them to survive hibernation. In terms of physiological resources this makes good sense, both for the females who produce late litters and for the resulting young.

Sun, before giving birth to two hoglets Emily and Bramwell

Ivy found with five babies beneath a portacabin

Mums and babies

It is not uncommon for nests of hedgehogs to be disturbed during the breeding season. This might result from the dismantling of a shed or decking. If at all possible, the nest should be left in situ and a cover positioned to provide protection from the elements. If the nest has been destroyed, or if it is wholly impractical to leave it where it is, the entire family of hedgehogs needs to be relocated. On several occasions I have had to bring a mother with babies to my garden, or a friend's garden. I placed them in a nest box in the garden, with food and water close by.

It is important to keep interference to a minimum. In every instance, the mother continued to feed her babies. She quickly found the food and water provided and chose to remain in the nest box I had placed her in. However, this does not always happen. I recall a mother and five babies that had been found under a portacabin that was being moved on a building site. Richard Jackson, a member of staff at York RSPCA, drove the return journey from York to Bradford and, at my suggestion, delivered them to a friend's garden where a specially made hedgehog house was waiting for them.

At dusk, the mother hedgehog, Ivy, could be seen moving each of her babies, one at a time, to another part of the garden where she created her own nest.

In August 2012 I cared for two albino siblings, one of which was female and named Sun. They were released to an area I monitor with cameras. In July 2014, Sun was spotted in her release area during daylight and I brought her in for observation.

She gave birth to two babies three days later. A camera I positioned inside the two-tiered 8' x 3' enclosure showed that she left the next box each evening at dusk. She walked down the ramp that led to the upper nest box, and did not return for up to eight hours. The babies were alone during this period. This agrees with other accounts that suggest that mother hedgehogs do not return to the nest during the night, but spend the time foraging.

Sadly, Sun died, unexpectedly, when the babies were seventeen days old. The vet suspected an internal infection linked to giving birth. She had been feeding normally and returning to her babies before dawn. After Sun died I hand reared her babies and they were returned to the area where their mother had been living for the past two years.

The arrival of three babies, in July 2016, to a hedgehog named Mystery, under observation in my garden, came as quite a surprise. She lined her nest with grass from the garden, regularly visited the feeding station, usually a couple of hours before dusk, and reared her babies very well as the photos show.

One day old hoglets in nest: born to Mystery in my garden

Nine days old

Twelve days old

Twenty nine days old

It is generally best to leave mother and babies undisturbed, following the birth, in order not to upset the mother. However, in certain circumstances I think that occasional checking of hoglets is warranted, for instance, if the mother appears to not be completely healthy and there are doubts about her ability to feed her young. In these circumstances I tend to check the hoglets every few days, using thin disposable gloves so my scent is not left in the nest. In the case of Mystery and her babies, intermittent checking was carried out, mainly to ensure that another hedgehog had not entered the nest from the garden and ousted Mystery.

It was apparent that at 36 days old the hoglets, named Mollie, Mott and Pete, were leaving the nest to forage on their own. Small distinct foraging holes could be clearly seen on my lawn.

On August 13, I noticed Mollie flash past the back door on her own to the feeding station. Shortly afterwards Mystery appeared, flanked closely by Mott, then the smallest of the three babies. Two days later, when the hoglets were 38 days old, Pete (weighing only 283g) left the nest and took up residence in an upper nest box of an open enclosure, after climbing a ramp with a gradient of 1 in 1.4 (70% slope). The following night he returned to the nest to be with his mum and siblings. The night after he struck out again on his own and found an entirely different nest box in which to spend the night.

Table 3   Weights of Mystery's hoglets

Date	Age (days)	Wt (g) Pete M	Wt (g) Mollie F	Wt (g) Mott M
260716	18	118	113	105
050816	28	204	180	163
100816	33	229	206	205
130816	36	252	238 ticks: 1	232
150816	38	283 ticks: 1	247	245 ticks: 2
160816	39	287	229	245 ticks: 1
170816	40	282	249	-
180816	41	282 ticks: 1	251 ticks: 2	254 ticks: 2
190816	42	293	271 ticks: 1	244 ticks: 4
220816	45	338 ticks: 1	257	276
230816	46	351	351 ticks: 7	303 ticks: 10
240816	47	338 ticks: 3	250	303 ticks: 5
250816	48	342	265	305 ticks: 1
260816	49	354	258	338 ticks: 1
310816	54	375	281	358
070916	61	435	381	410

Pete (283g) managed to climb this ramp to reach the upper nest box

Mystery, shortly after giving birth (1075g)

Table 3 shows that Pete, one of the males, remained the heaviest for the first 61 days. It tends to be case that the difference in weight between hoglets when they are born, tends to be maintained throughout the growth period. Ticks infested the hoglets at times and were immediately removed (see p 74).

Mystery's babies will be released to the wildlife allotment garden that I manage, subject to the weather and their weights. On September 5 Mystery began to hibernate so her release has been deferred.

Small hole burrowed in lawn showing that the young hoglets were foraging at only 36 days old

Mollie (257g), Pete (338g), Mott (276g) 45 days

Caring for hoglets

Young hedgehogs that would normally still be with their mother are called hoglets. They generally weigh less than 250g. Whenever one or more hoglets appear in the open, during daylight hours, it can generally be assumed that something has happened to the mother. If the mother does not return to the nest, after spending the night foraging, the hoglets will venture out of the nest, driven by hunger.

Occasionally, when the hoglets are very tiny, they can be heard 'peeping' from their nest, producing a sound not unlike that of baby birds. If hoglets are 'peeping' but have not left their nest, it is best to wait for up to 24 hours to see if their mother will return. Once it has been established, beyond all measure of doubt, that the hoglets are without a mother and are in desperate need of help, it is important to act immediately.

First weigh all the hoglets, then mark using a small amount of nail varnish or emulsion paint on the spines to enable you to distinguish between the siblings. This is important when assessing and weighing them each day so you can identify whether one might be in trouble and losing weight.

Next place the hoglet(s) on a towel close to a vet recovery heat pad. Cover with another towel with no ragged edges as hoglets can become caught inside torn edges and have been known to be strangled.

If hoglets are mobile and eating on their own place a small, shallow dish of water in their box, as well as food. The type of food is dependent on how old they are. Hoglets from birth to around four weeks old can be fed Esbilac, which is a puppy milk substitute that contains all the ingredients necessary for normal growth. I have heard that goat's milk has also been used to successfully rear entire litters of hedgehogs.

I use a 1ml syringe to feed them, as this allows you to administer one drop at a time of the liquid and be able to quantify how much each one has taken. It is important to hold the hoglet upright, so that no liquid can trickle back down the trachea and flood the lungs, which might result in inhalation pneumonia.

How well they do depends partly on how long they have been deprived of their mother's milk before they are found and taken into care. It also depends on how old they were when they became separated from their mother.

I once took charge of four siblings that had become accidentally trapped in a barn. The farmer found three of them four days after closing the barn door and unintentionally excluding the mother hedgehog. The fourth hoglet was discovered a day later than the others.

All four survived and were released back to the wild. However, they were lucky. In most cases, hoglets that have been deprived of food, for any length of time, develop tummy upsets (diarrhoea) and loss of appetite. In these instances they are prescribed the oral antibiotic Baytril, which usually does the trick and acts quickly. Also, even small hoglets can occasionally suffer from a heavy burden of intestinal parasites such as lungworm (treated with Telmin) and tapeworm (treated with Panacur).

Hoglets need toileting for the first two weeks or so of life, until they are able to manage on their own. This can be achieved by simulating what the mother would do. A cotton wool bud, coated with Vaseline, is ideal for this purpose. Rubbed gently over the genitals it will simulate the action of the mother's tongue and cause the hoglet to wee and poo. The Vaseline is to make sure that the tender skin of the hoglet does not become sore during the process.

How can you tell how old hoglets are? If the eyes are still closed they are probably less than fourteen days old. The photographs and weights shown for Mystery's babies in 'Mums and babies' will give a good indication of age and stages of development.

How much weight can you expect hoglets to gain on a daily basis? Those born early in the year will normally gain around 10g a day, while those born later in the year can be expected to gain over 12g a day. It is important to weigh each animal daily as they can quickly take a turn for the worse and require treatment.

Figure 2 shows the daily weights recorded for eight siblings that came into my care. In each case the hoglet was healthy and gained weight at an acceptable rate.

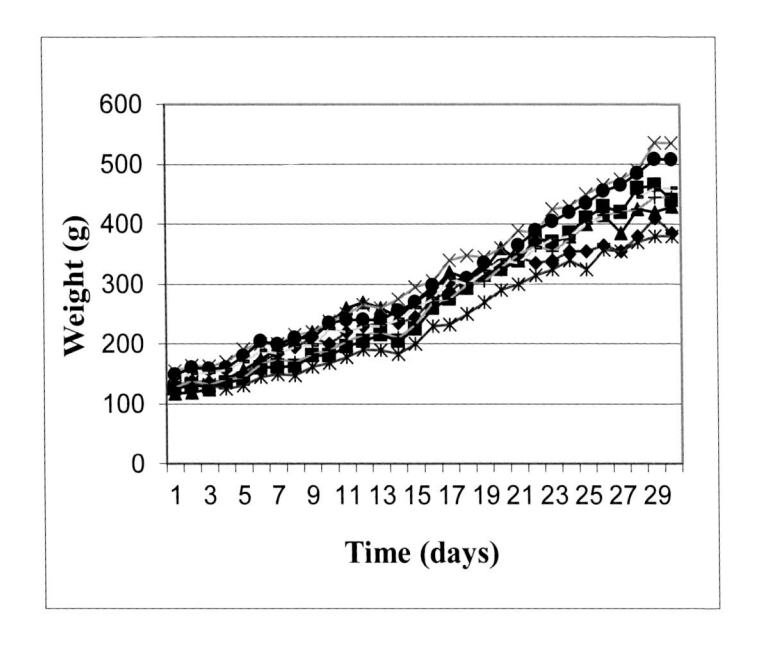

Figure 2 Growth curves for eight healthy siblings

Litter sizes can vary considerably. The highest number of siblings I cared for was eight. The photos show the difference in size that can exist between siblings at quite an early age.

Litter of four

Obvious size difference between these siblings

Male    Female

Case study: Rosy

Rosy has been chosen as a good example of a typical hoglet that has been found out in the day, away from the nest. Some hoglets are straightforward to deal with and only need feeding. Rosy did not fit this profile. She arrived on August 7, 2016. This is her story.

Day 1   110g Eating egg. Lost weight. Started Baytril.
Day 2   105g Not eating. Syringe fed lactose-free milk. Started Telmin and Panacur.
Day 3   102g Continued with syringe feeding: liquidised Chum meat loaf.
Day 4   100g Syringe feeding Esbilac. Very weak and listless. Looks beyond hope.
Day 5   104g Continued syringe feeding. Starting to eat raw egg mixed with Chum meat loaf.
Day 6   101g Syringe feeding Hill's Prescription Diet a/d . Starting to lap on own properly.
Day 7   98g Syringe feeding Hill's Prescription a/d.
Day 8   95g Eating Hill's Prescription a/d on own.
Day 9   108g Eating Hill's Prescription a/d on own.
Day 10 122g Eating Hill's Prescription a/d on own.
Day 11 132g Eating Hill's Prescription a/d on own.
Day 12 143g Ate two small dishes of Chum meat loaf
Day 13 159g, Day 14 179g, Day 15 189g, Day 16 215g, Day 17 228g, Day 18 237g, Day 19 260g, Day 20 279g
Rosy then went to outdoor enclosure until she gained enough weight to be released to her finding site.

Above left: Rosy on arrival, 110g

Above: Losing weight day 2

Left: Being syringe fed

Rosy keeping snug and warm on the heat pad while eating

Above: Rosy regains her appetite when presented with a dish of scrambled egg (no milk) that she promptly stands in.
Below: Rosy at 260g, about to be moved outside.

Sight impaired hedgehogs

During the first twenty years of caring for hedgehogs I saw only one with impaired vision. In the past six years I have seen about a dozen. In most cases eyes were present but appeared milky and failed to respond to treatment such as eye ointment. Although hedgehogs mostly depend on the senses of hearing and smell, sight is also important. It has been shown to help them to locate landmarks. This is only to be expected considering that hedgehogs are crepuscular, rather than nocturnal. This means that they venture forth around dusk and are active until dawn.

A hedgehog with impaired sight is vulnerable and not a suitable candidate for releasing back to the wild. There are two possible courses of action: Euthenase or release to an enclosed, fully supported garden. I opt for the latter. But how can you tell if a hedgehog has difficulty seeing? Close examination of the eye might reveal some damage, but if there is some residual vision the hedgehog will probably be able to avoid obstacles and detect changing light intensity. This would render return to the wild a distinct possibility. There are two criteria I use when assessing the eyesight of a hedgehog. One is the ability to be able to avoid objects placed randomly in their path. The other is observing whether the hedgehog emerges from its nest during the day.

Beacon enjoys the freedom to roam safely

Beacon finds the feeding station and water with ease

Philly by the back door

The latter might not indicate eyesight problems, as this abnormal behaviour can also be caused by a head injury. Regardless of the cause, any hedgehog that is otherwise healthy, but emerges in the day, cannot be released. Having established that a hedgehog needs to reside in a safe environment for the rest of its life, what constitutes an area large enough to roam? Ideally, as large as possible, but, faced with the scarcity of enclosed gardens where the owner is prepared to monitor the hedgehog and provide food and water every night, it is more realistic to look at what might be an acceptable minimum.

For the past few years I have had several long term residents in my enclosed garden. These included Roger (The Disappearing Hedgehog book) who had suffered a strimming injury to the head. He appeared to thrive well and exhibited normal behaviour with the exception of forays during the day. My garden is about 10m long and 3m wide. Dotted throughout the undergrowth are several hedgehog houses which serve as nest boxes. Three 'blind' hedgehogs were rehomed in my garden and I was able to observe their behaviour at close quarters. They were named Beacon, Philly and Lucy Lu, Beacon being the only male. Beacon was able to locate a female during the breeding season and to attempt to mate with her, albeit unsuccessfully. All three were able to find their nest box of choice, as well as food in the feeding station and water dishes in the garden. Beacon sometimes chose to sleep in the upper quarters of the two-tiered enclosure that also served as a feeding station. He was able to manoeuvre his way up the ramp with ease. Philly had a habit of waiting near the back door in the early evening, waiting for fresh food to be put out, while Lucy Lu chose to nest underneath the base of a wooden hutch. She also gathered straw and grass. Her efforts to line her nest with grass were admirable. They also provided some insight into the oft-heard tales about hedgehogs eating grass.

For several minutes at a time Lucy Lu would walk about on my lawn, in broad daylight, tearing up the grass with her teeth. When she had accumulated quite a mouthful she would carry it back and line her nest with it. I have since witnessed this behaviour in another female, Mystery, who spent a few days adding grass to her nest box, despite the fact that it was already filled to the brim with hay. She then promptly gave birth to three babies. This anecdotal evidence suggests that nest building behaviour is typical of hedgehogs when they are about to give birth. Interestingly enough, the grass gathering took place regardless of the plentiful nest material available, suggesting that this behaviour is innate.

Beacon, Philly and Lucy Lu lived in my garden for over three years. I am satisfied that this is a viable option for hedgehogs who, for whatever reason, cannot be returned to the wild.

Philly and Beacon meet up

Beacon attempts to mate Philly: unsuccessfully

Lucy Lu: always out in the morning

Lucy Lu gathering grass and using it to line her nest

Taking precautions to ensure that she does not lose the grass she has gathered, Lucy Lu has wrapped it between two leaves.

Why do ticks target hedgehogs?

Several years ago I was checking two adult hedgehogs that were being monitored in my garden, prior to release. They had chosen to nest in the same box and were lying close together, so close in fact that their spines were touching in several places. It struck me as strange that one of the hedgehogs was covered in ticks and the other had none at all. This made me wonder what might cause the hedgehog tick (*Ixodes hexagonus*) to infest some hedgehogs, but not others. The available literature seemed to suggest that it was all down to chance, but my observation suggested otherwise.

The hedgehog tick is not an ambushing tick like *Ixodes ricinus* that typically lies in wait beside paths frequented by deer, one of their hosts. By contrast, the hedgehog tick can often be found in nest boxes, carried there on a hedgehog. Once fully gorged a tick will detach itself and lie dormant until ready for its next blood meal. It will then attach itself to a hedgehog visiting the nest box. However, in the case I observed, this had not happened, only one of two hedgehogs becoming infested, despite the preponderance of ticks in the nest box.

I decided to investigate this and embarked on a study involving chemical analysis and a gas chromatography mass spectrometer (GC-MS) (Bunnell et al, 2011). The investigation looked at whether health status of the West European hedgehog (*Erinaceus europaeus*) is correlated with tick burden, and whether chemical cues linked to the health status of the host mediate attraction of the hedgehog tick, *Ixodes hexagonus*.

I conducted an ecological survey involving data collected over ten years from 226 wild hedgehogs that came to me for care, having been found out in the day. The study revealed a strong association between health status and tick burden of hedgehogs, with healthy animals being less likely to carry ticks than unhealthy ones. Behavioural choice tests demonstrated that ticks display a preference for the faecal odour from sick hedgehogs compared with healthy ones.

Chemical analysis of faecal odours using a GC-MS showed that, of seven volatile compounds identified in the odour profiles, only one substance, indole, was present in significantly higher amounts in sick hedgehogs compared with healthy ones. Ticks were also found to be attracted to indole when given the choice between indole and a solvent control. This implies that it is the attraction to faecal odour that causes higher tick burdens in sick hedgehogs.

Ticks might benefit from this preference by avoiding possible repulsion mechanisms of healthy hosts. The findings suggest that ticks potentially choose their host based on odour linked to the host's health status.

Why would ticks use faecal odour to locate their hedgehog host? As odour sources of hedgehogs are limited to anal glands and sweat glands under the feet, attraction of ticks from a distance may therefore be limited to footprints and faeces. As the hedgehog tick *I. hexagonus* appears to dwell in open vegetation, as well as in nests, faecal deposits could provide a cue that enables ticks to target their hedgehog hosts in areas where hedgehogs are foraging, and are either stationary or slow moving. Hedgehogs also tend to use similar routes while away from the nest, and are likely to pass by faeces deposited previously. Hence, hedgehog faecal deposits are a good place for ticks to encounter a passing host.

Faecal products can provide information regarding the physiological condition of a host, including general health. Ticks that target hosts with reduced immune status are likely to have a selective advantage over those that target hosts irrespective of host resistance.

Sick hedgehogs were shown to be infested by a significantly higher number of ticks than healthy ones. The attraction of ticks to the faeces of sick hedgehogs, rather than to the faeces of healthy ones, suggests that ticks selectively target sick hedgehogs.

It is safe to assume that if a hedgehog arrives at a rescue centre, for treatment, the presence or absence of ticks will help to determine its state of health.

Above and left:

Ticks infesting a hedgehog in large numbers.

Below:
Ticks in a nest box

Hibernation

I am in the process of analysing hibernation data from my own rescue centre, spanning 26 years, and those from other rescue centres across the UK. Currently I am able to provide a snapshot of hibernation patterns chosen at random. The triggers for the onset of hibernation have been well documented. These include change in day length and fall in ambient temperature.

It is important to state that hibernation in the hedgehog is not essential for survival. If it is warm enough they might not hibernate at all, something that is witnessed on the North Island of New Zealand. Following the introduction of the European Hedgehog to the country in the 1880s, by the Canterbury Acclimatisation Society, the animals have done very well indeed, to the extent that they are now widely regarded as a pest due to their tendency to predate ground-nesting birds and their eggs.

In the winter of 2012 minimum temperatures dropped to as low as -14° C in North Yorkshire, where York Hedgehog Rescue is based. In these conditions a 200g hedgehog was found hibernating while another, weighing around 200g, was discovered eating bird seed in the snow in someone's garden.

Sex	Hibernation onset date	Weight (g)	No. of periodic arousals	Final emergence date	Weight (g)
F	081112	682	4	230213	514
F	231112	662	11	180213	554
M	281112	1134	14	270213	984
M	291112	1070	1	051212	1211
M	081212	1220	5	180213	1520
F	091013	467	2	301113	573
M	111013	1226	2	100114	1114
M	111013	1324	3	270114	1232
F	261013	648	3	050114	533
F	281013	1091	4	110114	1015
F	021113	879	9	110314	648
M	051113	991	0	221213	609
M	131113	536	3	221213	651
F	141113	826	0	100114	678
F	231113	842	1	260114	793
F	051213	565	3	010114	543
F	131114	660	5	170215	647
F	040315	654	0	080315	590
F	211115	737	4	260216	597
F	170116	610	1	200216	640

Table 4 Hibernation: onset and emergence

There is one important difference between hedgehogs and other native UK mammals that hibernate, such as the dormouse. Hedgehogs do not always hibernate continuously and instead display what is known as periodic arousal.

Table 4 shows that the number of times an individual animal emerges from hibernation and enters it again, varies tremendously. Some hedgehogs emerge from hibernation heavier than they were at the onset. This is due to emerging several times and eating during these episodes of periodic arousal. Data from twenty randomly chosen animals are depicted in the table and have not been subjected to statistical analysis. However, it is immediately evident that the weight and date at which they first start to hibernate vary considerably.

In 2015, three of my hedgehogs hibernated at very different times and weights: George began on January 1 at 1055g in the garden. Surray started on January 13 at 668g, also in the garden, while Zola went into hibernation on January 16 at 563g in my insulated shed. All hedgehogs had access to food and water at all times. One hedgehog in particular demonstrated his own trend. Snow, the albino male that I released to my monitored area in August 2012, displayed the following pattern:

Onset: Oct 4 2012   Emergence: Apr 8 2013
Onset: Sep 18 2013   Emergence: Mar 12 2014
Onset: Sep 28 2014   Emergence: Feb 26 2015

Weight / size relationship: A new index

One important consideration, when releasing hedgehogs as winter approaches, is their prospect of surviving hibernation. I conducted a study where I established a system for taking certain measurements in order to determine whether a hedgehog had a good chance of survival (Bunnell, 2002).

When assessing an animal's nutritional status and chances of survival, post-release, the relationship between weight and size was found to be important, rather than weight per se. Two girth measurements were taken, at right angles to each other, on a tightly curled hedgehog. By dividing the latitudinal circumference (A) by the longitudinal circumference (B) an index is arrived at. A minimum value for this index, in a healthy hedgehog with a good prognosis following release, was found to be 0.80.

In addition, the analysis of data collected from 168 hedgehogs, between April 1998 and December 2000, demonstrated that a minimum weight of 450 – 500g is not an acceptable criterion on which to base the release of rehabilitated hedgehogs when the onset of cold weather is imminent. Analysis of my data suggested a recommended minimum weight for hedgehogs, in the autumn, to be 650g (with an associated index of 0.80 or more). I actually opt to release at 800+g immediately prior to winter.

Hedgehog hibernating showing clearly that no part of the face or feet are visible. This ensures that they are not exposed to the cold air during the winter months.

Healthy hedgehog showing nice rounded end

Two malnourished hedgehogs with tapered end

A 	B

The latitudinal circumference (A) divided by the longitudinal circumference (B) gives the index.
This should be no less than 0.80 in a healthy hedgehog.

The borderline hedgehog

It is well documented that the hedgehog, once a familiar sight in our parks and gardens, has suffered a massive decline in numbers in the UK over the past few decades.

Members of the public, and hundreds of hedgehog and wildlife rescue centres, play an important role in rehabilitating those animals in need and returning them to the wild when possible.

When winter approaches, decisions have to be taken by hedgehog rescue centres regarding when to bring hedgehogs in for treatment and care. Hedgehogs found out in the day are obviously in need of help but there are those individuals, seen out at night, that raise questions as to the fate that should befall them: whether to leave them where they are, in the hope that they will hibernate if necessary, or whether to bring them inside.

Both options have advantages and disadvantages. Hedgehogs taken into rescue centres tend to remain there until the following spring, with few exceptions, sometimes resulting in stress due to being held in captivity. However, those with insufficient body mass, if left to fend for themselves, might not survive the winter.

The guidelines for deciding when to remove a hedgehog from the wild, when it is out at night, are often based on the weight of the animal alone. This is tenuous at best and disastrous at worst. A previously delivered minimum pre-hibernation weight of 450g, necessary for a hedgehog to survive hibernation, has been superseded by my recommendation of 650g (coupled with a rounded end) (Bunnell T. 2002).

But this is not the whole story. Just to confound the issue there is the 'borderline hedgehog'. I decided to coin this term after someone brought me a hedgehog in early autumn, 2015. She had found it the previous evening in a park in York. Knowing that hedgehogs should be a certain weight/size to be able to survive hibernation, she contacted me for advice. Not having any scales to weigh it (another problem faced by the public in these circumstances) she brought it to me. I found it to weigh 550g, with an almost rounded end, and three small ticks that I removed.

So far so good. What to do next? Decision time. The temperature was set to fall in York in the near future but given the weight/size ratio of the hog, and the fact that it still had time to reach 650g, it would probably be fine. And there we have it: the word 'probably'. The borderline weight implied that the hedgehog was on the threshold of being able to survive a period of hibernation. However, the presence of three ticks suggested that the hedgehog was not in perfect health, as outlined in the section about why ticks target hedgehogs. I decided that it was not worth taking a chance and the hedgehog was taken into care for the winter months until it had gained enough weight. It was subsequently released exactly where it was found in the park.

When deciding whether or not to remove a hedgehog from the wild, when it is out at night in the autumn, it is imperative that the overall health status of the animal is taken into consideration, as well as the ambient and predicted weather conditions.

Borderline hedgehog showing an almost perfectly rounded end but with slight signs of a taper

Post-release monitoring

As previously mentioned, whenever possible hedgehogs are returned to their finding site. When circumstances dictate that this is not possible, suitable alternatives need to be sought.

In York I am fortunate to have access to a large green area within the city walls, with adjoining green corridors, which is not accessible to the public. For six years I have been monitoring the hedgehogs that live there, including those released by me.

Using unique identification for each hedgehog, I have been able to determine whether hedgehogs relocated to the area remain there and, if so, for how long. In addition, it has allowed me an insight into when the onset of hibernation occurred for certain hedgehogs, coupled with the date of emergence.

The greater majority of hedgehogs that I released to the area, following treatment at my rescue centre, opted to remain there, regardless of where they had come from in the first place. This dispelled reports that hedgehogs tend not to remain in an area that is alien to them, and will try to find their way 'home'.

My study proved, categorically, that this did not happen, even though it would have been very easy for them to move away from the area along green corridors that travel across York.

Means of identification

For several years I have been marking hedgehogs prior to release. Using a small amount of emulsion paint, and taking care that it only goes on the spines, I make a unique mark which serves as identification for that particular hedgehog. When releasing animals to the green area I am monitoring, I ensure that the mark is identical on both sides of the hedgehog. This enables it to be picked up by the camera, regardless of the side that the animal presents as it walks past.

Emulsion marks can last up to a year, remaining visible to the Infra-Red camera, but are not so easily detectable by the naked eye in daylight. Should a hedgehog appear in the day, in need of help, identifying it could be somewhat problematic. With this in mind I decided to develop a new, fail-safe system. This involves removing the central wire from an electric cable and cutting small sections, about half the length of a hedgehog spine, to produce hollow 'tubes' that will serve as tags. Using tweezers I slide one tube onto a spine. Adding a drop of super glue to the top of the tube seals it in place. I often add more than one tag, or use different colour combinations.

A photographic record ensures that each hedgehog, so marked, can be easily identified at any time in the future. This technique has been found to be remarkably reliable, the tags remaining in place for well over a year.

Black emulsion paint still visible on albino one year after release, and visible on I-R cameras three years later.

Dane is marked with a unique emulsion mark and two blue plastic tags.

Malone is tagged with a brown and a blue plastic tag.

Two blue plastic tags are fixed in position on Moses using Superglue

Malone is released to the wild with his unique emulsion mark and plastic tag

Publications

Bunnell T. 1998. Susceptibility of juvenile hedgehogs to disease: Some observations. Imprint, Newsletter of the Yorkshire Mammal Group, No. 26.

Bunnell T. 2000. Tea Tree Antiseptic Cream: A new treatment for ringworm and sarcoptic mange in the hedgehog *Erinaceus europaeus.* Journal of the American Holistic Veterinary Association, September, Vol.19, No.2: 29-31.

Bunnell T. 2001. Treatment for ringworm and sarcoptic mange in the hedgehog *(Erinaceus europaeus).* The Rehabilitator, British Wildlife Council Newsletter 30: 5, January.

Bunnell T. 2001. An effective, harmless treatment for tick *(Ixodes hexagonus)* infestation in the hedgehog (Erinaceus europaeus). Journal of the American Holistic Veterinary Association, January, Vol 19, No.4: 25-26.

Bunnell T. 2001. The importance of fecal indices in assessing gastrointestinal parasite infestation and bacterial infection in the hedgehog (*Erinaceus europaeus*). Journal of Wildlife Rehabilitation, Vol.24 (2):13-17.

Bunnell T. 2001. The incidence of disease and injury in displaced wild hedgehogs (*Erinaceus europaeus*). Lutra, Vol. 44 (1):3-14.

Bunnell T. 2001. Hedgehog Rehabilitation (Letter). Journal of Wildlife Rehabilitation, Vol. 24 (4):3.

Bunnell T. 2002. Wild Hedgehogs. Mammal News, No. 130:7.

Bunnell T. 2002. The assessment of British hedgehog (*Erinaceus europaeus*) casualties on arrival and determination of optimum release weights using a new index. Journal of Wildlife Rehabilitation, Vol. 25, (4): 11-21.

Bunnell T. 2007. Hedgehogs in trouble. Yorkshire Wildlife Trust magazine: 12-13.

Bunnell T. 2009. Growth rate in early and late litters of the European hedgehog (*Erinaceus europaeus*). Lutra, Vol. 52 (1):15-22.

Toni Bunnell, Kerstin Hanisch, Joerg D. Hardege and Thomas Breithaupt. 2011. The fecal odor of sick hedgehogs (*Erinaceus europaeus*) mediates olfactory attraction of the tick *Ixodes hexagonus*. Journal of Chemical Ecology: Volume 37 (4): 340-347.

Speck S., Perseke L., Petney T., Skuballa J., Pfäffle M., Taraschewski H., Bunnell T., Essbauer S., Dobler G. 2013. Detection of Rickettsia helvetica in ticks collected from European hedgehogs (*Erinaceus europaeus Linnaeus*, 1758). Ticks and tick-borne diseases 4 (3): 222-226.

Many of these publications can be obtained in full: www.tonibunnell.co.uk

The Disappearing Hedgehog (book), Bunnell T. Available from: www.tonibunnell.com

Current research projects

I am currently analysing photographs collected over five years using remote Infra-Red Reconyx cameras. Long-term monitoring of hedgehog populations will hopefully add to our understanding of how hedgehogs behave, how far they travel following release, and how long they survive.

Data collected over the past 26 years are also being analysed in an attempt to shed more light on the factors that govern hibernation in hedgehogs.

Acknowledgements

I would like to thank all the staff at York RSPCA Animal Home, in particular Deputy Manager, Ruth McCabe and Marie Sandle, Sue Heathcote, Richard Jackson, Helen Limbert, Anne Stephenson and Geraldine Murgatroyd, for the invaluable support that they have shown for hedgehogs in need. Also, for the help and advice they have given me during the 26 years that I have been running York Hedgehog Rescue Centre.

The help and support provided by the staff at the Minster Veterinary Practice, York, led by vets Don McMillan and Mark Goodman, has also been much appreciated. The co-operation of the veterinary staff has helped to develop a greater knowledge base regarding how best to treat the hedgehog.

Thank you also for the friendship and support of so many hedgehog carers throughout the UK. There are too many to mention but, in particular, I would like to thank Dru Burdon, Lorraine Jackson, Emma Farley, Beryl Steadman, Amber Leigh Glossop, Jackie Coup, Jacqui Morrell, Julie Smith, Irene Thomson, Mary Pope, Joy Hunt, Kathy Weeks, Adrian Buchan, Elaine Acton-Whittle, Louise Sandles, Gill and Phil Barrett, and Julie Campbell (Walk for Wildlife and Woofers).

During the many years that I have been monitoring hedgehogs in a green area within the city walls of York, I have been given tremendous support by the gardeners who manage the estate. Whenever they have found hedgehogs out in the day they have contacted me immediately. In addition they have built hedgehog houses and made sure that food and water was available at all times. Thanks very much indeed to Martin Ball, Lee Waudby, Peter Carlill and Stephen Murray.

Regular communication between hedgehog rehabilitators enables us to keep up to date with the latest treatments and also to share valuable experiences, all of which help immensely when trying to reach our goal; namely to restore to full health as many sick hedgehogs as possible, with the aim of releasing them back to the wild. Hopefully this book will help to achieve this, while at the same time helping to reduce the rate of decline of hedgehog numbers in the UK.

The Author

Toni Bunnell has had a lifelong passion for animals and their welfare, with an academic career that includes a degree in zoology, an MSc in gastroenterology and a doctorate in polecat behaviour. She lectured in physiology for sixteen years at Hull University, conducting research in the fields of chemical ecology and hedgehog conservation, welfare and ecology. This resulted in several publications most of which are available on her website: www.tonibunnell.co.uk

In 1990 Toni began running, single-handedly, York Hedgehog Rescue Centre, taking in hedgehogs that were in need of care. These included sick or injured animals and orphaned babies. The knowledge gained during the years of working on a daily basis with hedgehogs helped greatly with the writing of The Disappearing Hedgehog book and also this book.

Toni conducts research into various aspects of hedgehog ecology, including hibernation and litter sizes. She also carries out long-term monitoring of hedgehog populations in York. She has appeared many times on radio and television, including Countryfile, Love Your Garden and, in 2015, The One Show. In 2017 Toni was awarded an Animal Action Award by IFAWUK, at the House of Lords, for outstanding achievement in the field of hedgehog care and conservation.